ENERGY SECTOR STANDARD OF THE PEOPLE'S REPUBLIC OF CHINA
中华人民共和国能源行业标准

Specification for Engineering Geological Investigation of Pumped Storage Power Stations

抽水蓄能电站工程地质勘察规程

NB/T 10073-2018

Chief Development Department: China Renewable Energy Engineering Institute
Approval Department: National Energy Administration of the People's Republic of China
Implementation Date: March 1, 2019

China Water & Power Press
中国水利水电出版社
Beijing 2024

All rights reserved. No part of this publication may be reproduced, stored in a retrieval system, or transmitted in any form or by any means—electronic, mechanical, photocopying, recording or otherwise, without prior written permission of the publisher.

图书在版编目（CIP）数据

抽水蓄能电站工程地质勘察规程：NB/T 10073-2018 = Specification for Engineering Geological Investigation of Pumped Storage Power Stations (NB/T 10073-2018)：英文 / 国家能源局发布. -- 北京：中国水利水电出版社，2024. 10. -- ISBN 978-7-5226-2917-9

Ⅰ．TV743-65

中国国家版本馆CIP数据核字第2024PN5177号

ENERGY SECTOR STANDARD
OF THE PEOPLE'S REPUBLIC OF CHINA
中华人民共和国能源行业标准

Specification for Engineering Geological Investigation
of Pumped Storage Power Stations
抽水蓄能电站工程地质勘察规程
NB/T 10073-2018
（英文版）

Issued by National Energy Administration of the People's Republic of China
国家能源局　发布
Translation organized by China Renewable Energy Engineering Institute
水电水利规划设计总院　组织翻译
Published by China Water & Power Press
中国水利水电出版社　出版发行
　Tel: (+ 86 10) 68545888　68545874
　sales@mwr.gov.cn
　Account name: China Water & Power Press
　Address: No.1, Yuyuantan Nanlu, Haidian District, Beijing 100038, China
　http: //www.waterpub.com.cn
中国水利水电出版社微机排版中心　排版
北京中献拓方科技发展有限公司　印刷
184mm×260mm　16开本　3.75印张　119千字
2024年10月第1版　2024年10月第1次印刷
Price（定价）：￥620.00

Introduction

This English version is one of China's energy sector standard series in English. Its translation was organized by China Renewable Energy Engineering Institute authorized by National Energy Administration of the People's Republic of China in compliance with relevant procedures and stipulations. This English version was issued by National Energy Administration of the People's Republic of China in Announcement [2021] No. 6 dated December 22, 2021.

This version was translated from the Chinese Standard NB/T 10073-2018, *Specification for Engineering Geological Investigation of Pumped Storage Power Stations* published by China Water & Power Press. The copyright is reserved by National Energy Administration of the People's Republic of China. In the event of any discrepancy in the implementation, the Chinese version shall prevail.

Many thanks go to the staff from relevant standard development organizations and those who have provided generous assistance in the translation and review process.

For further improvement of the English version, any comments and suggestions are welcome and should be addressed to:

China Renewable Energy Engineering Institute
No. 2 Beixiaojie, Liupukang, Xicheng District, Beijing 100120, China
Website: www.creei.cn

Translating organizations:

China Renewable Energy Engineering Institute
POWERCHINA Beijing Engineering Corporation Limited

Translating staff:

GAO Yan   QI Wen   GUO Jie   ZHOU Fuqiang

Review panel members:

QIE Chunsheng   Senior English Translator

QIAO Peng   POWERCHINA Northwest Engineering Corporation Limited

LIU Xiaofen   POWERCHINA Zhongnan Engineering Corporation Limited

ZHANG Ming   Tsinghua University

YAN Wenjun	Army Academy of Armored Forces, PLA
LIU Qing	POWERCHINA Northwest Engineering Corporation Limited
ZHANG Jin	POWERCHINA Kunming Engineering Corporation Limited
PENG Peng	POWERCHINA Huadong Engineering Corporation Limited
ZHAO Xiaoping	POWERCHINA Chengdu Engineering Corporation Limited
WANG Huiming	China Renewable Energy Engineering Institute

National Energy Administration of the People's Republic of China

翻译出版说明

本译本为国家能源局委托水电水利规划设计总院按照有关程序和规定，统一组织翻译的能源行业标准英文版系列译本之一。2021年12月22日，国家能源局以 2021 年第 6 号公告予以公布。

本译本是根据中国水利水电出版社出版的《抽水蓄能电站工程地质勘察规程》NB/T 10073—2018 翻译的，著作权归国家能源局所有。在使用过程中，如出现异议，以中文版为准。

本译本在翻译和审核过程中，本标准编制单位及编制组有关成员给予了积极协助。

为不断提高本译本的质量，欢迎使用者提出意见和建议，并反馈给水电水利规划设计总院。

地址：北京市西城区六铺炕北小街 2 号
邮编：100120
网址：www.creei.cn

本译本翻译单位：水电水利规划设计总院
　　　　　　　　中国电建集团北京勘测设计研究院有限公司
本译本翻译人员：高　燕　齐　文　郭　洁　周福强
本译本审核人员：

　　郄春生　英语高级翻译

　　乔　鹏　中国电建集团西北勘测设计研究院有限公司

　　刘小芬　中国电建集团中南勘测设计研究院有限公司

　　张　明　清华大学

　　闫文军　中国人民解放军陆军装甲兵学院

　　柳　青　中国电建集团西北勘测设计研究院有限公司

　　张　进　中国电建集团昆明勘测设计研究院有限公司

　　彭　鹏　中国电建集团华东勘测设计研究院有限公司

　　赵小平　中国电建集团成都勘测设计研究院有限公司

　　王惠明　水电水利规划设计总院

国家能源局

Announcement of National Energy Administration of the People's Republic of China [2018] No. 12

According to the requirements of Document GNJKJ [2009] No. 52 "Notice on Releasing the Energy Sector Standardization Administration Regulations (*tentative*) and detailed implementation rules issued by National Energy Administration of the People's Republic of China", 204 sector standards such as *Coal Mine Air-Cooling Adjustable-Speed Magnetic Coupling*, including 54 energy standards (NB), 8 petrochemical standards (NB/SH), and 142 petroleum standards (SY), are issued by National Energy Administration of the People's Republic of China after due review and approval.

Attachment: Directory of Sector Standards

National Energy Administration of the People's Republic of China

October 29, 2018

Attachment:

Directory of Sector Standards

Serial number	Standard No.	Title	Replaced standard No.	Adopted international standard No.	Approval date	Implementation date
…						
28	NB/T 10073-2018	Specification for Engineering Geological Investigation of Pumped Storage Power Stations			2018-10-29	2019-03-01
…						

Foreword

According to the requirements of Document GNKJ [2013] No. 235 issued by National Energy Administration of the People's Republic of China, "Notice on Releasing the Development and Revision Plan of the First Batch of Energy Sector Standards in 2013", and after extensive investigation and research, summarization of practical experience, and wide solicitation of opinions, the drafting group has prepared this specification.

The main technical contents of this specification include: engineering geological investigation for site selection, research on regional tectonic stability, engineering geological investigation of reservoir, engineering geological investigation of dam site, engineering geological investigation of water conveyance system, engineering geological investigation of underground powerhouse system, investigation of natural construction materials.

National Energy Administration of the People's Republic of China is in charge of the administration of this specification. China Renewable Energy Engineering Institute has proposed this specification and is responsible for its routine management. Energy Sector Standardization Technical Committee on Hydropower Investigation and Design is responsible for the explanation of specific technical contents. Comments and suggestions in the implementation of this specification shall be addressed to:

China Renewable Energy Engineering Institute
No. 2 Beixiaojie, Liupukang, Xicheng District, Beijing 100120, China

Chief development organizations:

China Renewable Energy Engineering Institute
POWERCHINA Beijing Engineering Corporation Limited

Chief drafting staff:

GONG Hailing	ZHANG Guobao	ZHAO Guogang	LI Jinfei
ZHANG Dongsheng	LI Yuanzhong	XIAO Haibo	MI Yingzhong
GUO Decun	PENG Shuojun	CHEN Hongyu	CHEN Tongfa
LI Jinmin	HAN Lijun	HOU Bo	LIU Zengjie
ZENG Sencai			

Review panel members:

PENG Tubiao	YANG Jian	WANG Huiming	GUO Yihua

WANG Wenyuan	SHAN Zhigang	LIU Chang	LI Xuezheng
CHEN Weidong	GUO Weixiang	ZHOU Zhifang	LI Sunquan
YI Zhijian	WANG Liangtai	ZHENG Kexun	LI Shisheng

Contents

1	**General Provisions**	1
2	**Terms**	2
3	**Engineering Geological Investigation for Site Selection**	3
3.1	General Requirements	3
3.2	General Survey of Potential Sites	3
3.3	Selection of Planned Sites	4
4	**Research on Regional Tectonic Stability**	7
4.1	Research Content	7
4.2	Research Method	7
4.3	Assessment on Regional Tectonic Stability	7
5	**Engineering Geological Investigation of Reservoir**	9
5.1	Investigation Content	9
5.2	Investigation Methods	10
5.3	Assessment on Reservoir Engineering Geology	13
6	**Engineering Geological Investigation of Dam Site**	17
6.1	Investigation Content	17
6.2	Investigation Methods	17
6.3	Assessment on Engineering Geology of Dam Site	19
7	**Engineering Geological Investigation of Water Conveyance System**	21
7.1	Investigation Content	21
7.2	Investigation Methods	21
7.3	Assessment on Engineering Geology of Water Conveyance System	23
8	**Engineering Geological Investigation of Underground Powerhouse System**	26
8.1	Investigation Content	26
8.2	Investigation Methods	26
8.3	Assessment on Engineering Geology of Underground Powerhouse System	27
9	**Investigation of Natural Construction Materials**	30
9.1	General Requirements	30
9.2	Content of Investigation on Natural Construction Materials in Reservoir and Underground Excavated Materials	30
9.3	Methods of Investigation on Natural Construction Materials in Reservoir and Underground Excavated Materials	30

9.4　Assessment on Engineering Geology of Natural Construction Materials in Reservoirs and Underground Excavated Materials ·· 31

Appendix A　Classification of Reservoir Leakage Patterns and Types ·· 32

Appendix B　Calculation of Reservoir Leakage ················· 33

Appendix C　Calculation of Dam Foundation Seepage ·········· 35

Appendix D　Curve Type Classification of Water Pressure Test in Borehole·· 38

Appendix E　Criteria to Determine the Minimum Thickness of Overlying Rock Mass on High Pressure Conduit ······ 40

Appendix F　Reduction Coefficient of External Water Pressure ·· 41

Appendix G　Calculation of Water Inrush in Underground Powerhouse Cavern ·· 42

Explanation of Wording in This Specification ······················ 44

List of Quoted Standards ·· 45

1 General Provisions

1.0.1 This specification is formulated with a view to standardizing the work content, methods and technical requirements for ensuring the quality of engineering geological investigation of pumped storage power stations.

1.0.2 This specification is applicable to the engineering geological investigation of pumped storage power stations.

1.0.3 The work depth of investigation for each stage shall comply with the current national standard GB 50287, *Code for Hydropower Engineering Geological Investigation*.

1.0.4 The engineering geological investigation of pumped storage power stations shall include:

  1 Comparison of alternative station sites and reservoir or dam sites in terms of engineering geological conditions.

  2 Reservoir leakage.

  3 Stability of the slopes inside and outside the reservoir.

  4 Routing of water conveyance system, and analysis of leakage and seepage stability of pressure tunnels.

  5 Selection of underground powerhouse system location and evaluation of surrounding rock stability of underground caverns.

  6 Investigation and utilization of natural construction materials within reservoir area.

1.0.5 In addition to this specification, the engineering geological investigation of pumped storage power stations shall comply with other current relevant standards of China.

2 Terms

2.0.1 water level fluctuation zone

range of reservoir water level variation in the pumping and generating cycle, generally referring to the slope zone between normal pool level and minimum operating level

2.0.2 thin watershed

portion of watershed divide near the normal pool level that has thin rock and soil mass and might cause reservoir leakage and slope failure

2.0.3 groundwater divide

ridge line of phreatic surface where groundwater seeps in different directions

2.0.4 aquiclude (relatively impermeable layer)

continuous rock and soil mass with a permeability rate or permeability coefficient less than a given value

2.0.5 reservoir bottom leakage

loss of water seeping from the reservoir in vertical direction

2.0.6 reservoir bank leakage

loss of water seeping from the reservoir in horizontal direction

2.0.7 dam foundation on slope

dam foundation built on a gully bottom with longitudinal slope greater than 15°

2.0.8 high pressure water test

in-situ pressure water test to determine the permeability and seepage stability of rock mass and the opening pressure of discontinuities under the action of high head

3 Engineering Geological Investigation for Site Selection

3.1 General Requirements

3.1.1 Site selection should proceed in two steps, i.e., general survey of potential sites and selection of planned sites.

3.1.2 Project sites shall be located in areas with favorable topographical and geological conditions, taking into account the horizontal length to head ratio.

3.1.3 Water retaining structures such as the dams of upper and lower reservoirs shall not be built on active faults.

3.1.4 Site selection should avoid areas with large-scale adverse geophysical phenomena.

3.1.5 When groundwater is proposed as the water source for makeup, the possibility and reliability of the groundwater as the makeup source shall be studied.

3.2 General Survey of Potential Sites

3.2.1 The engineering geological investigation for site survey shall include:

1. Topographical and geological conditions of sites.
2. Regional tectonic stability.
3. Major adverse geophysical phenomena and major engineering geological problems with sites.

3.2.2 The engineering geological investigation method for site survey should meet the following requirements:

1. The 1 : 50 000 to 1 : 10 000 topographic maps of the survey area should be used to identify the potential sites that meet the topographic conditions for arranging the upper reservoir, lower reservoir, and water conveyance and power generation system.
2. The regional geological data, seismic zoning data, and seismic safety evaluation results of projects in adjacent areas should be collected and analyzed to understand the regional geology and seismicity in the survey area.
3. The major adverse geophysical phenomena in the survey area should be studied based on the analysis of regional geological, seismic and hydrogeological data in the survey area and the interpretation of

available aerial photographs and satellite images.

3.2.3 The analysis and evaluation of engineering geological conditions for site survey shall include:

1 Suitability of topographical and geological conditions for the project.

2 Influence of regional tectonic stability on the project.

3 Influences of adverse geophysical phenomena and major engineering geological problems on the project.

3.3 Selection of Planned Sites

3.3.1 The engineering geological investigation for the selection of planned sites shall include:

1 Regional fault distribution and activity, seismicity and ground motion parameters of the site, to understand the effect of the regional fault on the project.

2 Development and distribution of major adverse geophysical phenomena in the site area, such as collapse, landslide and debris flow.

3 Basic engineering geological conditions of upper and lower reservoirs and dam sites, such as topography and geomorphy, stratum lithology, geological structure, weathering and relaxation characteristics of rock mass, and distribution of surface water and groundwater.

4 The thin watersheds around reservoirs, lower adjacent valleys, fault and fracture zones running through watersheds, ancient river course, karst development, distribution of springs and wells, shape of groundwater divide, etc.

5 Topography of bank slopes of upper and lower reservoirs as well as the stability of slopes inside and outside the reservoirs.

6 Engineering geological conditions along the route of the water conveyance and power generation system, such as topography and geomorphy, lithology, geological structure and groundwater, to understand the overlying rock mass thickness and surrounding rock stability of the underground caverns.

7 Engineering geological conditions of existing reservoirs and dams if they are used as the upper and lower reservoirs for the project.

8 Availability of natural construction materials at the site.

3.3.2 The engineering geological investigation methods for the selection of planned sites shall meet the following requirements:

1   Site reconnaissance should be conducted based on the latest regional geological data collected and the interpretations of aerial photographs and satellite images during the general site survey, to understand the development and distribution of major adverse geophysical phenomena such as collapse, landslides and debris flows at and around the planned sites.

2   Site investigation shall be conducted primarily by engineering geological mapping and aided with necessary geophysical exploration and light-duty exploration. The scope of engineering geological mapping shall cover the sites of reservoirs and dams and the route of the water conveyance and power generation system. The range of engineering geological mapping for a reservoir should cover the watershed or the adjacent valley. The scale of engineering geological mapping may be 1 : 10 000 to 1 : 5 000 for reservoir and dam areas, and 1 : 50 000 to 1 : 10 000 for the water conveyance and power generation system.

3   A representative exploratory profile shall be arranged respectively for the upper or lower reservoir dam site and the water conveyance and power generation system. For proposed near-term projects, boreholes shall be arranged in the main dam site area, the number of which should not be less than 3; when aided by geophysical exploration, 1 to 3 geophysical exploratory profiles should be arranged for the dam site.

4   Boreholes may be properly arranged in the potential leakage areas in the reservoir and dam sites, such as cols, thin watersheds, fault zones and karst development sites, to ascertain the buried depth of groundwater table; geophysical exploratory profiles may be arranged in deep valleys, shallow buried sections or thick overburden areas along the route of the water conveyance and power generation system.

5   The main rocks and soils of the sites should be subjected to lab test. The main rocks and soils lab tests and water quality analysis shall be carried out for the proposed near-term projects.

6   The sources of natural construction materials for the planned sites shall be surveyed.

3.3.3   The preliminary evaluation on the engineering geology of the planned site shall include:

1   Regional tectonic stability.

2　Influence of the major adverse geophysical phenomena on the project.

3　Analysis of the possibility of reservoir leakage, and preliminary geological recommendations on reservoir basin seepage control type.

4　Stability of the slopes inside and outside the upper and lower reservoirs.

5　Analysis of the major engineering geological problems for the upper reservoir, lower reservoir and dam site area, to assess the feasibility of reservoir formation and dam construction.

6　Preliminary analysis of the surrounding rock types and stability of the underground caverns of the water conveyance and power generation system of the planned site, and geological recommendations on selection of the route of the water conveyance and power generation system.

7　Distribution, reserves and quality of quarry and borrow areas and the availability of natural construction materials from the reservoir area.

3.3.4 The geological opinions on the planned site alternatives shall be put forward based on the major engineering geological problems with each planned site.

4 Research on Regional Tectonic Stability

4.1 Research Content

4.1.1 The research on regional tectonic stability shall include:

1 Regional tectonic setting.

2 Identification of active faults.

3 Seismic environment and seismic risk analysis.

4 Assessment on regional tectonic stability.

4.1.2 When the following conditions coexist, the amplification effect of seismic force should be considered:

1 The peak ground acceleration in project area is 0.1g or above.

2 The upper reservoir is located at the isolated peak.

3 The height difference between upper and lower reservoirs is greater than 400 m.

4.2 Research Method

4.2.1 The research method shall comply with the current standards of China GB 50287, *Code for Hydropower Engineering Geological Investigation*; and NB/T 35098, *Specification of Regional Tectonic Stability Investigation for Hydropower Projects*.

4.2.2 Seismic safety evaluation shall be carried out for the project if the dam height is over 100 m and the peak ground acceleration is 0.1g or above. The seismic safety evaluation method shall comply with the current national standard GB 17741, *Evaluation of Seismic Safety for Engineering Sites*. For other projects, the seismic ground motion parameters shall be determined in accordance with the current national standard GB 18306, *Seismic Ground Motion Parameter Zonation Map of China*.

4.2.3 The amplification effect of earthquake should be subjected to the mountain dynamic response test.

4.3 Assessment on Regional Tectonic Stability

4.3.1 The activity characteristics of the geotectonic unit where the project is located and its boundary faults shall be analyzed. The distribution of faults near the site and their activity shall be analyzed.

4.3.2 The distribution and activity of faults in the region shall be analyzed.

When active faults exist in the region, the spatial distribution characteristics of active faults shall be analyzed to evaluate the impacts on structures.

4.3.3 The seismogeological setting of the region shall be analyzed to determine the ground motion parameters.

4.3.4 Regional tectonic stability classification shall be carried out for the project site. The criterion for classification shall comply with the current sector standard NB/T 35098, *Specification of Regional Tectonic Stability Investigation for Hydropower Projects*.

4.3.5 Seismogeological hazards shall be evaluated.

5 Engineering Geological Investigation of Reservoir

5.1 Investigation Content

5.1.1 The reservoir engineering geological investigation shall comply with the current standards of China GB 50287, *Code for Hydropower Engineering Geological Investigation*; and DL/T 5336, *Technical Code of Reservoir Area Engineering Geological Investigation for Hydropower and Water Resources Project*.

5.1.2 The engineering geological investigation for reservoir leakage shall include:

1 Types and features of the topography and geomorphy in the reservoir areas, with focus on the shape, width and variation of cols and thin watersheds.

2 Lithology, thickness, distribution and weathering degree of each stratum in the reservoir area.

3 Distribution, scale, properties, extension, spatial distribution and permeability of major faults and densely fissured zones.

4 Hydrogeological structure of the rock and soil mass at the bottom or around the reservoir; the thickness and distribution characteristics of aquiclude, permeable layer, and permeable zone and their relationship with the reservoir water level; the permeability and seepage stability of permeable layers and permeable zones.

5 Groundwater type in the reservoir area; the recharge, runoff and discharge modes of groundwater; groundwater divide, groundwater level and their dynamic changes. Elevation, water volume and its dynamic changes of springs and wells.

6 Karst area investigation shall comply with the current sector standard NB/T 10075, *Specification for Karst Engineering Geological Investigation of Hydropower Projects*.

5.1.3 The engineering geological conditions of the impervious foundation of the reservoir basin shall be investigated.

5.1.4 The engineering geological investigation on the stability of reservoir bank shall include:

1 Distribution and features of adverse geophysical phenomena.

2 Distribution characteristics of the weak discontinuities and soft rock

mass on bank slopes and their influence on slope stability.

3   Rock and soil mass properties of bank slope in water level fluctuation zone and their stability conditions under the effects of reservoir water level fluctuation.

4   Engineering geological conditions of the slopes inside and outside the reservoirs, especially the cols and the thin watersheds, the stability status of the reservoir banks and the signs of deformation.

5   Stability conditions of the excavated slope in reservoir areas.

5.1.5 The engineering geological investigation of solid runoff shall include:

1   Hydrological and meteorological data, human activities, vegetation development and the occurrence of historical solid runoff.

2   Topographical and geomorphological conditions and catchment area of debris flow valley forming region, transporting region and accumulating region, and the distribution, composition and volume of loose solid sources.

3   Engineering geological conditions of prevention and control structures.

5.1.6 When the sediment retaining dam is built on a sediment-laden river, the storage reservoir and sediment trapping reservoir shall be investigated.

5.1.7 When an existing reservoir is used, the engineering geological conditions such as leakage, bank stability, bank collapse and immersion shall be checked.

5.1.8 When natural lakes are used as reservoirs of pumped storage power stations, their causes of formation shall be ascertained and the engineering geological problems such as lake water leakage and bank stability shall be investigated. For natural lakes formed by debris barriers, the stability conditions of the debris barrier shall be ascertained.

5.1.9 The engineering geological investigation for reservoir makeup shall include:

1   Engineering geological conditions of makeup structures and routes.

2   When spring is used as the source for reservoir makeup, the genesis of spring and the conditions of regional groundwater recharge, runoff and discharge shall be studied, and the spring volumetric flow rate and its seasonal variation shall be observed.

5.2 Investigation Methods

5.2.1 Reservoir leakage investigation methods shall meet the following requirements:

1 The scale of engineering geological mapping may be 1 : 5 000 to 1 : 1 000, and the mapping shall include lower adjacent valleys and possible leakage paths.

2 Suitable geophysical exploration methods should be used to detect the location, spatial distribution and groundwater level of possible permeable zones, to provide a reference basis for borehole arrangement.

3 Exploratory profiles shall be arranged according to topographical and hydrogeological conditions, leakage types, and seepage control schemes, with focus on the col, thin watershed and possible leakage areas.

4 Boreholes shall be arranged in areas where leakage might occur, and the spacing of boreholes depends on the leakage pattern and range; the boreholes shall reach 15 m below the top of aquiclude or 20 m to 50 m below the groundwater level.

5 Borehole hydrogeology tests shall be carried out in layers and sections.

6 For wells, springs, boreholes and outcrop points of groundwater in exploratory adits, the long-term observation of discharge or groundwater level shall be carried out, and the observation duration shall not be less than one hydrological year.

7 In karst area, one or more methods such as electrical exploration, ground penetrating radar, seismic exploration, CT, integrated logging and geothermal method should be used for reservoir leakage investigation, which may measure water seepage field, water temperature field, hydrochemical field and isotope field, to comprehensively evaluate reservoir leakage conditions.

5.2.2 The reservoir bank stability investigation method shall meet the following requirements:

1 The scale of engineering geological mapping may be 1 : 5 000 to 1 : 1 000, and the mapping shall include the slopes inside and outside the reservoir.

2 The layout of exploration work shall be carried out on the basis of geological mapping, and no less than 2 exploratory lines shall be arranged for unstable and potentially unstable slopes; the main exploratory profiles shall be arranged parallel to the possible sliding direction of the potentially unstable bank slopes with consideration

of the arrangement of the permeability profiles. The exploration methods are dominated with boreholes, adits or shafts, and should be arranged based on detecting concealed geological interface, thickness of potentially unstable bank slope and groundwater level by comprehensive geophysical exploration method.

3   The number of control boreholes in major exploratory profile should not be less than 3, and the boreholes shall extend 10 m to 20 m into the stable rock and soil mass.

4   When the excavation slopes of the reservoir banks are high or the geological conditions are complex, the exploratory adits should be arranged.

5   Sampling and the physical and mechanical property test or in-situ test for main rock and soil layers, weak interlayers, potential sliding surfaces or sliding surfaces which are critical to the bank slope stability shall be carried out in accordance with the current sector standard DL/T 5337, *Technical Code for Engineering Geological Investigation of Slope for Hydropower and Water Resources Project*.

6   Boreholes shall be subjected to groundwater level observation and hydrogeological test.

7   Necessary monitoring may be conducted for potentially unstable mass.

5.2.3   The foundation investigation method for the impervious layer of reservoir basin shall meet the following requirements:

1   The foundation of the impervious layer may be investigated in combination with the stability of reservoir bank, leakage of reservoir and the quarry and borrow area in reservoir primarily by drilling, pit exploration and well exploration, and the depth shall reach the relatively intact rock mass.

2   Samples shall be taken for lab test, and in-situ test may also be conducted.

5.2.4   The solid runoff investigation methods shall meet the following requirements:

1   The scale of engineering geological mapping may be 1 : 10 000 to 1 : 2 000 for solid runoff forming region, transporting region and accumulating region, and 1 : 2 000 to 1 : 500 for prevention and control structures.

2   Geophysical exploration, pit exploration, well exploration, and borehole drilling should be arranged in the provenance areas and the prevention and control structures, and the spacing of exploratory points and depth of boreholes should be determined according to specific conditions.

3   Physical and mechanical properties of the representative solid sources and deposits shall be tested.

4   When solid runoff has a great impact on the project, a special study should be carried out.

5.2.5   The investigation methods for existing reservoirs shall meet the following requirements:

1   The engineering geological data of the reservoir shall be collected and analyzed, and the engineering geological conditions shall be checked.

2   The corresponding exploration and test shall be arranged as required. Special engineering geological investigation shall also be carried out for renovation and extension projects.

5.2.6   Corresponding investigation work shall be arranged for the dedicated energy storage reservoir and sediment trapping reservoir on a sediment-laden river.

5.2.7   The natural lakes used as reservoirs of pumped storage power stations shall be investigated accordingly. Special engineering geological investigation shall be carried out for the debris barrier.

5.2.8   The investigation method for reservoir makeup project shall meet the following requirements:

1   The scale of engineering geological mapping may be 1 : 10 000 to 1 : 2 000 for makeup routes, and 1 : 1 000 to 1 : 500 for makeup structures.

2   The exploration should mainly use boreholes and pits, and the depth should reach the principal bearing stratum.

3   When spring water is used as the source for makeup, hydrogeological mapping shall be carried out, long-term observation of water level and flow rate of springs shall be carried out, and observation time shall not be less than three hydrological years.

5.3   Assessment on Reservoir Engineering Geology

5.3.1   The reservoir leakage is classified according to the hydrogeological structure, leakage pattern and boundary conditions of the reservoir area. The classification of reservoir leakage patterns and types shall be in accordance with

Appendix A of this specification. The reservoir leakage under natural conditions should be preliminarily estimated, and may be calculated in accordance with Appendix B of this specification.

5.3.2　The reservoir with one of the following conditions may be assessed as leakage free:

1　When there is no lower adjacent valley around the reservoir, and the reservoir is still the groundwater discharge base level in the project area after reservoir impoundment.

2　When the lowest water level of the groundwater divide around the reservoir is higher than the normal pool level of the reservoir.

3　When there is continuous aquiclude in the reservoir basin below the normal pool level.

5.3.3　A reservoir with one of the following conditions may be assessed as leakage prone:

1　There is no aquiclude between the reservoir and lower adjacent valley, and there is no groundwater divide or the groundwater divide is lower than the normal pool level.

2　There are large-scale fault-fractured zones or densely jointed zones extending beyond the reservoir, forming groundwater troughs lower than the normal storage level.

3　There are outward leakage passages such as ancient river courses, ancient weathered crusts, ancient erosion surfaces, and mine adits.

5.3.4　The assessment criterion for reservoir leakage in karst areas shall comply with the current sector standard NB/T 10075, *Specification for Karst Engineering Geological Investigation of Hydropower Projects*.

5.3.5　The influence of groundwater divide change on reservoir leakage after excavation of reservoir basins shall be analyzed.

5.3.6　The influence of reservoir leakage on underground structures should be evaluated, including water conveyance system and underground powerhouse. The change of groundwater level around the reservoir after impoundment and its influence on the surrounding hydrogeological environment should be predicted.

5.3.7　The permeability stability of thin watershed rock mass shall be evaluated according to the permeability characteristics.

5.3.8　The reservoir perimeter shall be zoned in terms of hydrogeology to determine the leakage type of each zone. Recommendations shall be put forward on the pattern, range and depth of reservoir seepage control according to the conditions and quantity of the reservoir bottom leakage and reservoir bank leakage.

5.3.9 Recommendations on the layout of hydrogeological monitoring network shall be put forward.

5.3.10 The reservoir bank stability shall be evaluated in accordance with the current sector standard DL/T 5337, *Technical Code for Engineering Geological Investigation of Slope for Hydropower and Water Resources Project*.

5.3.11 The stability of slopes in water level fluctuation zone shall be evaluated under hydrodynamic pressure or freeze-thaw action, and recommendations on treatment measures shall be put forward.

5.3.12 The slope stability shall be evaluated according to the natural slope topography, rock mass structure, rock weathering, relaxation, geological structure and possible reservoir leakage, and recommendations on treatment measures shall be put forward.

5.3.13 The slope stability shall be evaluated according to the rock mass structure and discontinuities combination of the excavated slopes in the reservoir, and recommendations on treatment measures shall be put forward.

5.3.14 The evaluation of solid runoff shall include:

1 Estimate the total provenance of solid runoff and the possible maximum volume into the reservoir.

2 Evaluate the influence of solid runoff on the projects and put forward recommendations on comprehensive treatment measures.

3 Evaluate the engineering geological conditions of prevention and control structures.

5.3.15 Engineering geological evaluation shall be carried out for the energy storage reservoir and sediment trapping reservoir on a sediment-laden river, respectively.

5.3.16 The uneven deformation of foundation of impervious body shall be evaluated according to the outcrop range and mechanical properties of foundation rock and soil layers, and recommended values of physical and mechanical parameters such as bearing capacity and deformation modulus of foundation of impervious body and recommendations on treatment measures shall be put forward.

5.3.17 The evaluation of makeup works shall include:

1   Evaluate the bearing capacity of the foundations of the makeup structures and the stability of the surrounding rock or slope according to the topography, lithology, physical and mechanical properties of rock and soil layer of the makeup structures and routes.

2   When a spring is used as the source for makeup, the regional hydrogeological features, groundwater recharge, runoff and discharge, water level and volumetric flow rate of spring shall be analyzed, and the possibility and reliability of spring as the supply source shall be evaluated.

5.3.18 The recommended values for physical and mechanical parameters of rock and soil mass and discontinuities in the reservoir area shall be put forward.

6 Engineering Geological Investigation of Dam Site

6.1 Investigation Content

6.1.1 Engineering geological investigation of dam site shall comply with the current standards of China GB 50287, *Code for Hydropower Engineering Geological Investigation*; DL/T 5414, *Code for Dam-site Project Geological Investigation of Hydropower and Water Resources*; and NB/T 10075, *Specification for Karst Engineering Geological Investigation of Hydropower Projects*.

6.1.2 The engineering geological investigation of dam foundation on slope shall include:

1. The valley topographic regularity, stratum lithology, geological structure, weathering degree of rock mass, development features of discontinuities, and physical and mechanical properties of dam foundation.

2. The development features, composition, physical and mechanical properties and leakage deformation features of potential sliding surfaces that affect the stability against shallow and deep sliding of dam foundation.

3. The engineering geological properties of foundation rock mass and development features of adverse rock and soil mass, and shear strength of interface between dam and foundation.

4. When setting up retaining walls, their foundations shall be investigated.

6.1.3 For dams around the reservoir which are constructed by excavation and fill, the engineering geological investigation of the dam site shall be carried out in combination with the reservoir area and quarry and borrow areas.

6.1.4 For the reservoirs with sediment retaining dams, the engineering geological investigation of sediment retaining dams and sediment flushing structures shall be carried out.

6.1.5 When existing reservoirs are used as upper and lower reservoirs, the engineering geological conditions of the dam shall be checked. When the dam needs renovation or extension, special engineering geological investigation shall be carried out.

6.2 Investigation Methods

6.2.1 The scale of engineering geological mapping may be 1 : 2 000 to

1 : 500. The mapping shall cover the sites of water retaining structures and the areas affecting the project.

6.2.2 Appropriate geophysical exploration methods may be selected according to the topographical and geological conditions of the site, taking into account the engineering geological problems.

6.2.3 Exploration arrangement shall meet the following requirements:

1 Each alternative dam site and dam axis shall be arranged with a primary exploratory profile. For the representative dam site with a dam height of 70 m or above and for the alternative dam sites with complex engineering geological conditions, secondary exploratory profiles should be added in the upstream and downstream of the primary exploratory profile.

2 The boreholes, adits and shafts shall be arranged for the main exploratory lines along dam axis and plinth. The spacing of exploratory points on the primary exploratory profiles should be determined depending on dam type. The borehole depth of the impervious curtain lines shall be 15 m below the top boundary of the aquiclude or 20 m to 50 m below the groundwater level.

3 Exploratory profiles shall be arranged along the maximum gradient of dam foundation on slope and should be investigated with boreholes, shafts or adits.

4 When a retaining wall is set for the dam foundation on slope, the exploratory profile should be arranged along the axis of the retaining wall.

5 The investigation of dams around the reservoir which are constructed by excavation and fill shall be combined with the investigation of the quarry and borrow areas, bank slopes and foundation of impervious body.

6 Geological data for the dam site of the existing reservoir shall be collected and reviewed, and exploration work shall be arranged when necessary.

6.2.4 The geotechnical test shall be suitable to the investigation stage, and be conducted in lab or in situ. The shear strength test of interface between dam and foundation should be carried out for dam foundation on slope.

6.2.5 The hydrogeological test shall meet the following requirements:

1 The hydrogeological test for the dam site area shall be conducted by investigation stages, considering the project scale, dam type, and

complexity of hydrogeological and engineering geological conditions.

2 The seepage deformation test may be carried out in the fault-fractured zone and weak interlayer in dam foundation as required.

3 The permeable layer and aquiclude shall be identified. In areas with complex hydrogeological conditions, groundwater connectivity test may be conducted as required.

6.2.6 During investigation, hydrogeological dynamics of borehole water level, spring volumetric flow rate, etc., shall be observed, and the observation duration shall not be less than one hydrological year.

6.2.7 The investigation methods for dam site in karst area shall comply with the current sector standard NB/T 10075, *Specification for Karst Engineering Geological Investigation of Hydropower Projects*.

6.3 Assessment on Engineering Geology of Dam Site

6.3.1 Geological opinions for alternative dam sites, dam types and dam axis shall be put forward according to the engineering geological conditions, hydrogeological conditions and major engineering geological problems.

6.3.2 The engineering geological assessment of foundation rock and soil mass shall comply with the current sector standard DL/T 5414, *Code for Dam-Site Project Geological Investigation of Hydropower and Water Resources*.

6.3.3 The leakage type of dam foundation shall be defined, the leakage quantity of dam foundation shall be calculated and the seepage stability of dam foundation shall be assessed. The dam foundation seepage can be calculated according to Appendix C of this specification.

6.3.4 The assessment of dam foundation on slope shall be conducted based on the topography, lithology, geological structure, rock mass structure, hydrogeological conditions, and bank slope structure. The overall stability of dam foundation slope and sliding stability of dam shall be evaluated, and geological recommendations for dam foundation treatment shall be proposed.

6.3.5 When the retaining wall is set for the dam foundation on slope, the engineering geological assessment shall be carried out for the bearing capacity and sliding stability conditions of the retaining wall foundation.

6.3.6 The foundation stability of dams around the reservoir which are constructed by excavation and fill shall be analyzed and evaluated in combination with the leakage of reservoir area and stability of reservoir bank.

6.3.7 The foundation stability and leakage features of the sediment retaining dam shall be evaluated and geological recommendations for leakage control and treatment shall be put forward.

6.3.8 When an existing reservoir is used, the composition and quality of dam materials shall be analyzed, and the physical and mechanical parameters and treatment recommendations for dam body and dam foundation shall be proposed.

7 Engineering Geological Investigation of Water Conveyance System

7.1 Investigation Content

7.1.1 The engineering geological investigation of water conveyance system shall comply with the current standards of China GB 50287, *Code for Hydropower Engineering Geological Investigation*; and DL/T 5415, *Technical Code for Underground Project Geological Investigation of Hydropower and Water Resources*.

7.1.2 The investigation for selected high pressure conduits shall include:

1 Thickness of overlying and lateral rock mass.

2 Magnitude and direction of rock mass geostress at different locations.

3 Hydraulic fracturing critical pressure and permeability of surrounding rock mass of different rock types.

7.1.3 When existing reservoirs or natural lakes are used, special engineering geological investigation shall be carried out for rock plugs and benches at the inlets/outlets of the conduits.

7.2 Investigation Methods

7.2.1 The scale of engineering geological mapping should be 1 : 5 000 to 1 : 2 000, and the mapping shall include alternative routes. When the topographical and geological conditions are complex, the longitudinal and transverse geological profiles should be measured, and the scale may be 1 : 2 000 or 1 : 1 000. The geological mapping scale for the inlet/outlet areas should be 1 : 2 000 to 1 : 500.

7.2.2 Geophysical exploratory profiles should be arranged along the routes to detect the development of concealed adverse geological phenomena.

7.2.3 Boreholes and adits shall be arranged at the conduit inlet/outlet areas in the upper and lower reservoirs, and the depth of adits shall meet the requirements for slope stability evaluation.

7.2.4 The investigation method for the rock plugs and benches at the conduit inlets/outlets shall meet the following requirements:

1 The scale of engineering geological mapping should not be less than 1:200.

2 Specialized joint and fissure statistics shall be carried out.

3 Boreholes shall be arranged and subjected to water pressure test. Borehole televiewer and inter-borehole geophysical CT should be used.

4 The physical and mechanical properties of rock samples shall be tested.

7.2.5 Boreholes shall be arranged at gate shafts, surge shafts, tunnel sections underneath gullies and shallow buried sections. Boreholes may be arranged at the headrace upper horizontal conduit section and the tailrace tunnel section according to the topographical and geological conditions and the rock cover above the tunnel. The depth of boreholes should be 10 m to 30 m below the tunnel bottom elevation. The water pressure test shall be carried out in boreholes, and long-term observation shall be conducted for groundwater level. The observation duration shall not be less than one hydrological year.

7.2.6 Boreholes shall be arranged along the high pressure conduits. The exploratory adit and boreholes for the high pressure bifurcation section shall be arranged in conjunction with the exploration of underground powerhouse.

7.2.7 The investigation method for the rock mass geostress surrounding the high pressure conduits shall meet the following requirements:

1 The magnitude and direction of rock mass geostress shall be measured in boreholes for high pressure conduits and high pressure bifurcation section, and the test method should be hydraulic fracturing. The geostress test shall comply with the current sector standard DL/T 5367, *Code for Rock Mass Stress Measurements of Hydroelectric and Water Conservancy Engineering*.

2 The regression analysis of initial geostress field in rock mass should be carried out.

7.2.8 The investigation method for permeability and hydraulic fracturing resistance of rock masses surrounding high pressure conduits shall meet the following requirements:

1 High pressure water test shall be carried out in the test section of complete rock mass, relatively complete rock mass or fissured rock mass in the high pressure conduit and high pressure bifurcation to test the permeability, permeability stability and hydraulic fracturing critical pressure of rock mass under the action of high water head. The high pressure water test shall comply with the current sector standard NB/T 35113, *Specification for Water Pressure Test in Borehole of Hydropower Projects*. The pressure-flow (P-Q) relationship shall be analyzed, and the borehole water pressure test curves should be classified in accordance with Appendix D of this specification.

2　　The maximum test pressure in the boreholes shall not be less than 1.2 times the maximum internal water pressure of the high pressure conduit at the specific location.

7.2.9　Representative rock samples shall be taken from exploratory adits and boreholes for physical and mechanical property test of the rock. Rock mass deformation test may be carried out in exploratory adits.

7.2.10　The measurement of ground temperature, harmful gas and radioactive content shall comply with the current sector standard DL/T 5415, *Technical Code for Underground Project Geological Investigation of Hydropower and Water Resources*.

7.3　Assessment on Engineering Geology of Water Conveyance System

7.3.1　Geological opinions on route selection shall be proposed according to the engineering geological conditions and main engineering geological problems with each route in water conveyance system.

7.3.2　The classification and stability evaluation of surrounding rocks shall comply with the current standards of China GB 50287, *Code for Hydropower Engineering Geological Investigation*; and DL/T 5415, *Technical Code for Underground Project Geological Investigation of Hydropower and Water Resources*.

7.3.3　The stability of inlet/outlet foundations and slopes shall be evaluated.

7.3.4　The stability evaluation of rock plugs and benches at the inlets/outlets shall include:

1　　Evaluation of topographical and geological conditions.

2　　Classification of rock mass structure and quality.

3　　Evaluation of the permeability and stability of rock plugs and benches under water pressure, and the stability of rock mass surrounding rock plugs after blasting.

7.3.5　The recommendations on the bifurcation location and the lining pattern of high pressure conduits shall be proposed according to the engineering geological conditions of rocks surrounding high pressure conduits.

7.3.6　The engineering geological evaluation of surrounding rock stability of reinforced concrete lined high pressure conduits shall include:

1　　Stability evaluation of mountain uplift. The minimum thickness of rock cover above the high pressure conduits shall be determined in accordance with Appendix E of this specification.

2 Evaluation of stability of surrounding rock against hydraulic fracturing. The high pressure conduits shall satisfy the requirement that the hydrostatic pressure in the tunnel is less than the minimum principal stress of surrounding rocks, and shall have a certain safety margin. The relationship between hydrostatic pressure and minimum principal stress of surrounding rocks can be calculated by the following formula:

$$F_1 h_s \gamma_w \leq \sigma_3 \qquad (7.3.6)$$

where

F_1 is the safety factor, generally 1.2 to 1.5;

h_s is the hydrostatic head in tunnel (m);

γ_w is the bulk weight of water (N/cm^3);

σ_3 is the minimum principal stress of the surrounding rock (MPa).

3 Evaluation of seepage stability of surrounding rocks. Reinforced concrete lined high pressure conduits should be used in Class I and Class II rock mass impermeable or slightly permeable, or in the rock mass that has the permeability less than 1.0 Lu after high pressure consolidation grouting and shall meet the seepage stability requirements.

7.3.7 The engineering geological evaluation of the surrounding rock stability of steel lined high pressure conduit shall include:

1 The ability of the surrounding rock to resist radial deformation shall be evaluated according to the unit elastic resistance coefficient, which should be determined by formula and engineering analogy, or by in-situ test. The elastic resistance coefficient per unit of the surrounding rock can be calculated by the following formula:

$$K_0 = \frac{E}{100(1+\mu)} \qquad (7.3.7)$$

where

K_0 is the unit elastic resistance coefficient of surrounding rocks (MPa/cm);

E is the elastic modulus of surrounding rocks (MPa);

μ is the Poisson's ratio of surrounding rocks.

2 According to the integrity and permeability of the overlying rock mass and the groundwater level, the external water pressure shall be

calculated by the reduction of underground water level. The reduction factor for the external water pressure shall be determined in accordance with Appendix F of this specification.

3 Recommendations on drainage measures for high pressure conduit sections should be put forward.

7.3.8 The water inrush and mud inrush prediction, ground temperature, harmful gas and radioactivity evaluation for conveyance system caverns shall comply with the current sector standard DL/T 5415, *Technical Code for Underground Project Geological Investigation of Hydropower and Water Resources*.

7.3.9 Recommendations for establishing hydrogeological monitoring network during operation period shall be put forward.

8 Engineering Geological Investigation of Underground Powerhouse System

8.1 Investigation Content

8.1.1 The engineering geological investigation of the underground powerhouse system shall comply with the current standards of China GB 50287, *Code for Hydropower Engineering Geological Investigation*; and DL/T 5415, *Technical Code for Underground Project Geological Investigation of Hydropower and Water Resources*.

8.1.2 The engineering geological investigation for the selected underground powerhouse site shall include:

1. Stratum lithology and geological structure.
2. Physical and mechanical properties of surrounding rocks.
3. Rock mass geostress.
4. Hydrogeological conditions.
5. Ground temperature, harmful gas and radioactivity.

8.2 Investigation Methods

8.2.1 The scale of engineering geological mapping should be 1 : 2 000 to 1 : 1 000. The mapping shall include the underground powerhouse caverns and ground switchyard.

8.2.2 The underground powerhouse exploration shall be conducted primarily by adit, in which boreholes/shafts and geophysical exploration shall be arranged. The exploration arrangement shall meet the following requirements:

1. Exploratory adits should be located on the side of tailrace tunnel of water conveyance system. If the topographic conditions permit, the length of the tunnel may be shortened or prolonged toward the underground powerhouse from the side of water conveyance system. The adit shall run through the location of high pressure bifurcation.

2. The adit portal should be arranged at a stable slope. When the portal is located upstream of lower reservoir dam, its elevation should be higher than the normal pool level. The elevation of adit bottom should be 30 m to 50 m higher than the powerhouse crown with a cross section not less than 2.2 m × 2.5 m, and the bottom slope should enable drainage by gravity.

3. Branch adits shall be arranged along the axis of the underground

powerhouse with an extension of no less than 50 m outside of the powerhouse end walls.

4   Structural tracing branch adits may be arranged inside exploratory adit according to the development of geological structures.

5   Boreholes shall be arranged inside the exploratory adit along the powerhouse axis and at the bifurcation, and the borehole spacing shall not be greater than 50 m. The borehole shall reach 10 m to 30 m below the structure base.

6   When the dip angle of rock stratum is gentle, the lithology is complex or the weak discontinuities of gentle dip angle is developed, shafts should be used to reach a certain depth below powerhouse crown elevation.

7   The elastic wave test of rock mass shall be carried out on the wall of the adit, and the borehole televiewer and geophysical CT should be used in the boreholes along the powerhouse axis.

8.2.3   The engineering geological investigation for the entrance and exit of each cavern of underground powerhouse system such as the access tunnel, ventilation and safety tunnel, shall be based on borehole exploration, pit exploration and trenching, and adits may also be arranged.

8.2.4   Representative rock samples shall be taken from adits and boreholes for lab test of physical and mechanical properties, and in-situ rock mass test should be carried out inside the exploratory adit. Samples shall be taken for analyzing the groundwater quality.

8.2.5   The geostress test shall be carried out in adits or boreholes, and at least two test methods should be used. The geostress test shall comply with the current sector standard DL/T 5367, *Code for Rock Mass Stress Measurements of Hydroelectric and Water Conservancy Engineering*.

8.2.6   The long-term observation of groundwater dynamics shall be carried out using boreholes and adits, and the observation duration shall not be less than one hydrological year.

8.2.7   Ground temperature, harmful gas and radioactivity tests shall comply with the current sector standard DL/T 5415, *Technical Code for Underground Project Geological Investigation of Hydropower and Water Resources*.

8.3   Assessment on Engineering Geology of Underground Powerhouse System

8.3.1   Geological opinions on each alternative powerhouse system layout scheme shall be put forward according to the basic engineering geological conditions and main engineering geological problems.

8.3.2 The location of the underground powerhouse system shall meet the following requirements:

1. The location of the underground powerhouse cavern shall be determined first, and then other caverns.

2. The section with intact topography, moderate burial depth, hard lithology, good rock mass integrity, relatively simple fracture structures and hydrogeological conditions should be selected.

8.3.3 The selection of underground powerhouse axis shall meet the following requirements:

1. The direction of the axis shall be determined by comprehensive analysis of the development features of fracture structures and the geostress status that affect the stability of surrounding rocks of the caverns.

2. The powerhouse axis should have a large angle with the main geological structure direction, and the angle should not be less than 60°.

3. For high geostress zones, the angle between the powerhouse axis and the maximum principal stress of surrounding rocks should not be greater than 30°.

4. When the requirements for maximum stress direction and main geological structure direction cannot be satisfied at the same time, the geostress should be considered for the high geostress zones, and the main geological structure direction should be considered for low or medium geostress zone.

8.3.4 The surrounding rocks of underground powerhouses shall be classified in accordance with the current standards of China GB 50287, *Code for Hydropower Engineering Geological Investigation*; and DL/T 5415, *Technical Code for Underground Project Geological Investigation of Hydropower and Water Resources*, the values of physical and mechanical parameters of rock mass and discontinuities shall be recommended, the possibility of rock burst shall be predicted, and the stability of surrounding rocks evaluated.

8.3.5 The hydrogeological conditions of the plant area shall be evaluated, the water inrush in underground powerhouse caverns shall be predicted, and drainage measures shall be recommended. Water inrush in underground powerhouse caverns should be calculated in accordance with Appendix G of this specification. The corrosiveness of groundwater shall be evaluated.

8.3.6 The evaluation of ground temperature, harmful gas and radioactivity shall comply with the current sector standard DL/T 5415, *Technical Code for Underground Project Geological Investigation of Hydropower and Water Resources*.

9 Investigation of Natural Construction Materials

9.1 General Requirements

9.1.1 The excavated materials of the upper reservoir, lower reservoir and caverns shall be fully utilized, and the cut-fill balance should be achieved.

9.1.2 When the reserves and quality of excavated materials inside the reservoir cannot meet the project needs, the sources of construction materials outside the reservoir shall be considered. The investigation of construction materials outside the reservoir shall comply with the current standards of China GB 50287, *Code for Hydropower Engineering Geological Investigation*; and DL/T 5388, *Code of Natural Building Material Investigation for Hydropower and Water Resources Project*.

9.1.3 The minimum reserve coefficient of the reservoir basin quarry, considering cut-fill balance, may be 1.2. When the reserve coefficient is less than 1.5, additional quarry should be considered.

9.2 Content of Investigation on Natural Construction Materials in Reservoir and Underground Excavated Materials

Investigation for reservoir natural construction materials and underground excavated materials shall include:

1. The lithology, geological structure, minerals and their chemical composition, development degree of discontinuities and fillers, etc.

2. The overburden thickness, weathering features, karst development degree and cave-fissure fillers, etc.

3. Weathering features, distribution of weak interlayers, karst development degree and mud intercalation.

4. Physical and mechanical properties of rocks.

5. Reserves, exploitation and transportation conditions.

9.3 Methods of Investigation on Natural Construction Materials in Reservoir and Underground Excavated Materials

9.3.1 The survey, exploration and test of natural construction materials may be carried out in conjunction with the investigation of reservoir and dam area, and the exploration and test data of the reservoir and dam area may be utilized. The investigation method shall comply with the current sector standard DL/T 5388, *Code of Natural Building Material Investigation for Hydropower and Water Resources Project*.

9.3.2 The survey, exploration and test of excavated materials may be carried out in conjunction with the investigation of caverns, and the investigation data of caverns may be fully utilized. The investigation method shall comply with the current sector standard DL/T 5388, *Code of Natural Building Material Investigation for Hydropower and Water Resources Project*.

9.4 Assessment on Engineering Geology of Natural Construction Materials in Reservoirs and Underground Excavated Materials

9.4.1 The quality and availability of natural construction materials shall be assessed by zone based on the genesis, lithology or material composition, stratification, weathering degree, and lab and field test results of natural construction materials.

9.4.2 The total reserves and available quantity, sub-zone reserves and available quantity, and the quantity and quality for different filling zones of natural construction materials shall be analyzed.

9.4.3 The reserve calculation and quality assessment shall be in accordance with the current sector standard DL/T 5388, *Code of Natural Building Material Investigation for Hydropower and Water Resources Project*.

9.4.4 The exploitation and transportation conditions of natural construction materials shall be analyzed and assessed.

9.4.5 The quality and utilization rate of underground excavated materials shall be assessed according to their lithology, weathering degree and test data.

9.4.6 In order to make full use of natural construction materials, cut-fill balance should be achieved. The use of completely and strongly weathered materials as dam materials should be studied.

Appendix A Classification of Reservoir Leakage Patterns and Types

A.0.1 The classification of reservoir leakage patterns by medium shall comply with Table A.0.1.

Table A.0.1 Classification of reservoir leakage patterns by medium

Leakage patterns		Main characteristics
Pore type		Large leakage from loose overburden or completely and strongly weathered layers
Fissure type	Structural zone leakage	Leakage from permeable layer and fault zone connecting inside and outside of the reservoir
	Bedrock fissure leakage	Leakage in the fissure-developed rock mass
Conduit type		Leakage in karst areas

A.0.2 The classification of reservoir leakage types by path shall comply with Table A.0.2.

Table A.0.2 Classification of reservoir leakage types by path

Leakage types	Main characteristics
Horizontal leakage	Leakage in horizontal direction along the thin watersheds, faults and fissure dense zone
Vertical leakage	Leakage along the vertical direction at the bottom of the suspended reservoir and the deep groundwater

Appendix B Calculation of Reservoir Leakage

B.0.1 When the horizontal leakage dominates (Figure B.0.1), the volumetric flow rate can be calculated by the following formula. For multiple permeable layers or a layer of different permeable zones, the weighted average value can be taken for permeability coefficient.

$$Q = KA\left(\frac{H_1 - H_2}{L}\right) \tag{B.0.1}$$

where

Q is the volumetric flow rate (m³/d);

K is the permeability coefficient (m/d);

A is the cross-sectional area of flow (m²);

H_1 is the reservoir normal pool level (m);

H_2 is the outcrop point elevation (m);

L is the leakage path length (m).

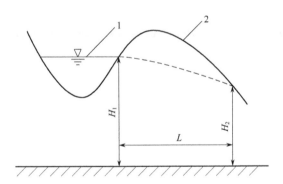

Key

1 normal pool level

2 terrain

Figure B.0.1 Calculation of horizontal leakage volumetric flow rate

B.0.2 When the vertical leakage dominates (Figure B.0.2), the volumetric flow rate can be calculated by the following formula. For multiple permeable layers or a layer of different permeable zones, the weighted average value can be taken for permeability coefficient.

$$Q = KF\left(\frac{H+l}{l}\right) \tag{B.0.2}$$

where

F is the reservoir leakage area (m²);

H is the height from reservoir normal pool level to reservoir bottom (m);

l is the height from reservoir bottom to aquiclude ceiling or groundwater table (m).

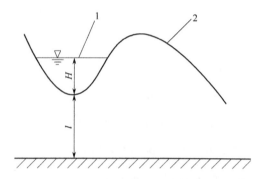

Key

1 normal pool level

2 terrain

Figure B.0.2 Calculation of vertical leakage volumetric flow rate

Appendix C   Calculation of Dam Foundation Seepage

C.1   Calculation of Dam Foundation Seepage

C.1.1   When the homogeneous permeable layer is infinite deep and the dam bottom is a plane, the volumetric flow rate of the homogeneous semi-infinite dam foundation seepage (Figure C.1.1) can be calculated by the following formulae:

$$Q = KBHq_r \qquad (C.1.1\text{-}1)$$

$$q_r = \frac{1}{\pi}\operatorname{arcsh}\frac{y}{b} \qquad (C.1.1\text{-}2)$$

where

Q is the dam foundation volumetric flow rate (m³/d);

B is the dam length (m);

K is the permeability coefficient (m/d);

H is the head from upstream to downstream (m);

q_r is the calculation seepage;

y is the calculation depth (m);

b is the half of the dam bottom width (m).

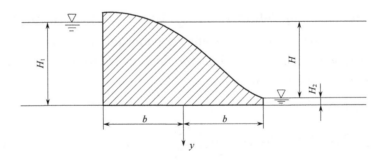

Figure C.1.1   Volumetric flow rate of homogeneous semi-infinite dam foundation seepage

C.1.2   When the homogeneous permeable layer has a finite depth ($M \leq 2b$) and the bottom protection is a plane, the volumetric flow rate of the homogeneous finite depth dam foundation seepage (Figure C.1.2) can be calculated by the following formula:

$$Q = KBH \frac{M}{2b+M} \qquad (C.1.2)$$

where

    Q is the dam foundation volumetric flow rate (m³/d);

    K is the permeability coefficient (m/d);

    B is the dam length (m);

    H is the head from upstream to downstream (m);

    $2b$ is the dam bottom width (m);

    M is the permeable layer thickness (m).

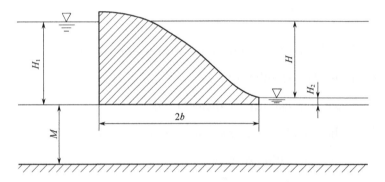

Figure C.1.2 Volumetric flow rate of homogeneous finite deep dam foundation seepage

C.2 Calculation of Volumetric Flow Rate of Bypass Seepage

Under the condition that the homogeneous permeable layers and the bottom slabs are nearly horizontal, the volumetric flow rate of bypass seepage can be calculated by the following formulae:

Free surface flow:

$$Q = 0.366 K (H_1 - H_2)(H_1 + H_2) \lg \frac{B}{r_0} \qquad (C.2.1)$$

Pressure flow:

$$Q = 0.732 K (H_1 - H_2) M \lg \frac{B}{r_0} \qquad (C.2.2)$$

where

    Q is the volumetric flow rate (m³/d);

    K is the permeability coefficient (m/d);

H_1 is the distance from normal pool level to aquiclude;

H_2 is the distance from groundwater level to aquiclude before impoundment (m);

B is the width of bypass seepage zone (m) (Figure C.2.1), which can be obtained by dividing L (Figure C.2.2) by π. L is the distance from the point where the elevation of groundwater level is equal to the normal pool level to the river bank;

r_0 is the circular radius (m) around the seepage line at the dam body joint, which can be obtained by dividing the circumference of the contour of dam joint by π;

M is the thickness of confined aquifer (m).

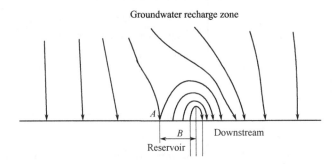

Figure C.2.1　Width of bypass seepage zone

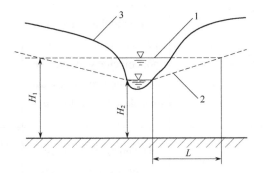

Key

1　normal pool level

2　groundwater level before impoundment

3　ground line

Figure C.2.2　Calculation diagram of bypass seepage zone

Appendix D Curve Type Classification of Water Pressure Test in Borehole

Table D Curve type classification of water pressure test in borehole

Type	P-Q curve	Curve characteristics	Interpretation
Type A (laminar flow)		The pressure rise curve is a straight line passing through the origin. The pressure drop curve coincides almost with the rise curve	The seepage regime is laminar flow, and the fissures remain unchanged during the test
Type B (turbulent flow)		The pressure rise curve is convex to Q axis, and the pressure drop curve coincides almost with the rise curve	The seepage regime is turbulent (known as all non-linear P-Q relations); the fissures remain unchanged during the test
Type C (expansion)		The pressure rise curve is convex to the P-axis, and the pressure drop curve coincides almost with the rise curve	The fissures change and permeability increases; this change is reversible and features elastic expansion*
Type D (scouring)		The pressure rise curve is convex to the P-axis. The pressure drop curve does not coincide with the rise curve. The two curves take on a closed loop clockwise	The fissures change and permeability increases; this change is irreversible and results mainly from the scouring and movement of fillers
Type E (filling)		The pressure rise curve is convex to the Q-axis. The pressure drop curve does not coincide with the rise curve. The two curves take on a closed loop counterclockwise	The fissures change and permeability decreases, resulting mainly from partial blockage of fissure, or by the semi-closed fissures being filled with water

38

Table D *(continued)*

Type	P-Q curve	Curve characteristics	Interpretation
Type F (strata uplift)		The pressure rise curve is convex to the Q-axis. The pressure drop curve does not coincide with the rise curve. The two curves take on an open loop counterclockwise	The fissures change and permeability decreases; a water body is formed due to the partial blockage, closed, semi-closed expansion or stratum uplift, which flows back when the pressure drops to a certain degree

*Translator's Annotation: This Table D in Appendix D is translated in accordance with Announcement [2021] No. 5, Attachment 3.2—Sector Standard Amendment Notification issued by National Energy Administration of the People's Republic of China on November 16, 2021.

Appendix E Criteria to Determine the Minimum Thickness of Overlying Rock Mass on High Pressure Conduit

The minimum thickness of overlying rock mass on high pressure conduits (Figure E.0.1), shall be determined according to the requirement that the hydrostatic pressure in the tunnel is less than the gravity of rock mass above the tunnel, and is calculated by the following formula:

$$C_{RM} = \frac{h_s \gamma_w F}{\gamma_r \cos \alpha} \qquad (E.0.1)$$

where

C_{RM} is the minimum thickness of overlying rock mass excluding completely and strongly weathered zones and strongly unloaded zone (m);

h_s is the hydrostatic head in tunnel (m);

γ_w is the bulk weight of water (kN/m³);

γ_r is the bulk weight of rock (kN/m³);

α is the valley bank slope angle (°), taken as $\alpha = 60°$ when $\alpha > 60°$;

F is the empirical coefficient, generally taken as 1.3 to 1.5.

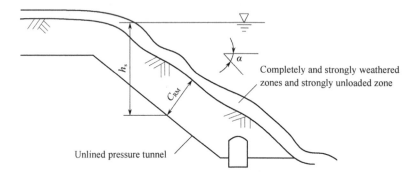

Figure E.0.1 Minimum thickness of overlying rock mass of high pressure conduits

Appendix F Reduction Coefficient of External Water Pressure

Table F Reduction coefficient of external water pressure

Level	Groundwater activity	Groundwater effect on the stability of surrounding rock	Reduction coefficient
1	Wall, dry or wet	None	0 - 0.20
2	Water seepage or water dripping from the discontinuities	The filler in the discontinuities, the shear strength of discontinuities, and the weak rock mass weakened	0.10 - 0.40
3	Severe dripping; a large amount of drip, water tearing or spraying along the discontinuities	The filler in the discontinuities argillized, the shear strength and the medium hard rock mass weakened	0.25 - 0.60
4	Severe dripping; a small amount of water inrush along the discontinuities	The filler in the discontinuities scoured, rock weathering accelerated, the weak zones such as faults weakened, argillized, swelled disintegrated and physical piping. The thin weak layer opened by seepage pressure	0.40 - 0.80
5	Severe gushing water; a large amount of water inrush from faults and other weak zones	The filler in the discontinuities scoured out, rock mass separated, faults and weak zones with certain thickness opened by seepage pressure, and surrounding rock collapsed	0.65 - 1.00

NOTE When there is good connectivity between the karst underground rivers, caves and surface water, the reduction coefficient is taken as 1.00.

Appendix G Calculation of Water Inrush in Underground Powerhouse Cavern

G.0.1 For fan-shaped aquifers, the water inrush of underground powerhouse caverns can be calculated by the following formulae:

Phreatic water formula:

$$Q = \frac{K(b_1 - b_2)}{\ln b_1 - \ln b_2} \cdot \frac{(h_1^2 - h_2^2)}{2L} \tag{G.0.1-1}$$

Confined water formula:

$$Q = \frac{KM(b_1 - b_2)(H_1 - H_2)}{(\ln b_1 - \ln b_2)L} \tag{G.0.1-2}$$

where

- Q is the water inrush in underground cavern (m³/d);
- M is the average thickness of confined aquifer in the fan-shaped section (m);
- K is the average permeability coefficient of aquifer (m/d);
- h_1, H_1, b_1 are the phreatic aquifer thickness, confined water level and width of the upstream calculation section (m);
- h_2, H_2, b_2 are the phreatic aquifer thickness, confined water level and width of the downstream calculation section (m);
- L is the average distance between upstream and downstream sections (m).

G.0.2 When a cavern runs through a phreatic aquifer, the maximum water inrush in the cavern can be predicted by the following formulae:

1) Goodman formula:

$$Q_0 = L \frac{2\pi K \cdot H}{\ln \frac{4H}{d}} \tag{G.0.2-1}$$

where

- Q_0 is the maximum water inrush of tunnel section through aquifer (m³/d);
- K is the permeability coefficient of aquifer (m/d);

- H is the distance from the hydrostatic level to the equivalent circle center of the tunnel cross section (m);
- d is the equivalent circle diameter of the tunnel cross section (m);
- L is the length of the tunnel section passing through an aquifer (m).

2) Satoh formula:

$$q_0 = \frac{2\pi \cdot m \cdot K \cdot h_2}{\ln\left[\tan\frac{\pi(2h_2 - r_0)}{4h_c} \cot\frac{\pi \cdot r_0}{4h_c}\right]} \quad \text{(G.0.2-2)}$$

where

- q_0 is the maximum water inrush per unit length of tunnel through aquifer [m³/(s · m)];
- m is the conversion coefficient, generally taken as 0.86;
- K is the permeability coefficient of aquifer (m/d);
- h_2 is the distance from the hydrostatic level to the equivalent circle center of the tunnel cross section (m);
- r_0 is the equivalent circle radius of the tunnel cross section (m);
- h_c is the thickness of aquifer (m).

G.0.3 When a cavern runs through one or more water catchments, the normal water inrush of the cavern can be predicted by the groundwater runoff modulus method and calculated by the following formulae:

$$Q_s = M \cdot A \quad \text{(G.0.3-1)}$$

$$M = \frac{Q'}{F} \quad \text{(G.0.3-2)}$$

where

- Q_s is the normal water inflow of tunnel section through aquifer (m³/d);
- M is the underground runoff modulus [m³/(d·km²)];
- Q' is the groundwater recharge flow rate from rivers or gravity springs (m³/d), taking the flow rate in low-flow period;
- F is the surface catchment area with flow rate equivalent to Q' of the surface water or gravity spring (km²);
- A is the catchment area of tunnel section through aquifer (km²).

Explanation of Wording in This Specification

1. Words used for different degrees of strictness are explained as follows in order to mark the differences in executing the requirements in this specification.

    1) Words denoting a very strict or mandatory requirement:

       "Must" is used for affirmation; "must not" for negation.

    2) Words denoting a strict requirement under normal conditions:

       "Shall" is used for affirmation; "shall not" for negation.

    3) Words denoting a permission of a slight choice or an indication of the most suitable choice when conditions permit:

       "Should" is used for affirmation; "should not" for negation.

    4) "May" is used to express the option available, sometimes with the conditional permit.

2. "Shall meet the requirements of..." or "shall comply with..." is used in this specification to indicate that it is necessary to comply with the requirements stipulated in other relative standards and specification.

List of Quoted Standards

GB 17741,	*Evaluation of Seismic Safety for Engineering Sites*
GB 18306,	*Seismic Ground Motion Parameter Zonation Map of China*
GB 50287,	*Code for Hydropower Engineering Geological Investigation*
NB/T 10075,	*Specification for Karst Engineering Geological Investigation of Hydropower Projects*
NB/T 35098,	*Specification of Regional Tectonic Stability Investigation for Hydropower Projects*
NB/T 35113,	*Specification for Water Pressure Test in Borehole of Hydropower Projects*
DL/T 5336,	*Technical Code of Reservoir Area Engineering Geological Investigation for Hydropower and Water Resources Project*
DL/T 5337,	*Technical Code for Engineering Geological Investigation of Slope for Hydropower and Water Resources Project*
DL/T 5367,	*Code for Rock Mass Stress Measurements of Hydroelectric and Water Conservancy Engineering*
DL/T 5388,	*Code of Natural Building Material Investigation for Hydropower and Water Resources Project*
DL/T 5414,	*Code for Dam-Site Project Geological Investigation of Hydropower and Water Resources*
DL/T 5415,	*Technical Code for Underground Project Geological Investigation of Hydropower and Water Resources*